YOUR KNOWLEDGE HAS VALUE

Nana Yaw Addo-Korie

LNG - A Review of Current and Future Markets

GRIN Verlag

Bibliografische Information der Deutschen Nationalbibliothek:

Die Deutsche Bibliothek verzeichnet diese Publikation in der Deutschen National-
bibliografie; detaillierte bibliografische Daten sind im Internet über http://dnb.d-
nb.de/ abrufbar.

Imprint:

Copyright © 2011 GRIN Verlag GmbH
Druck und Bindung: Books on Demand GmbH, Norderstedt Germany
ISBN: 978-3-656-10569-5

This book at GRIN:

http://www.grin.com/en/e-book/187167/lng-a-review-of-current-and-future-markets

GRIN - Your knowledge has value

Der GRIN Verlag publiziert seit 1998 wissenschaftliche Arbeiten von Studenten, Hochschullehrern und anderen Akademikern als eBook und gedrucktes Buch. Die Verlagswebsite www.grin.com ist die ideale Plattform zur Veröffentlichung von Hausarbeiten, Abschlussarbeiten, wissenschaftlichen Aufsätzen, Dissertationen und Fachbüchern.

Visit us on the internet:

http://www.grin.com/

http://www.facebook.com/grincom

http://www.twitter.com/grin_com

LNG – A REVIEW OF CURRENT AND FUTURE MARKETS

ABBREVIATIONS

LNG   -   Liquefied Natural Gas

OPEC -   Organisation of Petroleum Exporting Countries

CO_2 -   Carbon Dioxide

CO_2 e-   Carbon Dioxide equivalents

GHG -   Greenhouse Gases

EIA   -   Energy Information Administration

CO_2 -   Carbon Dioxide

GTI   -   Gas Technology Institute

BP   -   British Petroleum

UAE -   United Arab Emirates

BCM -   Billion cubic meters

TCF -   Trillion Cubic Feet

OECD- Organisation for Economic Co-operation and Development

IEA   -   International Energy Agency

MMT - Million metric tons

ABSTRACT

The LNG industry is experiencing strong growth in its current market- posting an impressive
297.63 bcm in LNG imports during 2010(BP 2011: 29). However, it is still a budding industry
because Russia and Iran- owners of the largest gas reserves in the world are still fledgling
LNG exporters (Economides and Wood 2009). According to Kumar et al. (2011) LNG is also a
clean substitute over petrol and diesel. Moreover the Asian tigers like Japan, China and India
are still growing their economies (ExxonMobil 2010: 8). Thus, the Asian tigers' LNG demand,
the eco-nature of LNG, coupled with Russia and Iran's LNG exports - can create a more
robust LNG market in the future. This paper examines the existing and future market of LNG
– in the light of major LNG players in each region.

1.   INTRODUCTION

The world is now frowning on "dirty" energy. Clearly, this is shown in the way natural gas
contends with traditional fuels, from a "clean" and "efficiency" outlook (Economides and
Wood 2009: 1). In fact, if the world's economy keeps growing –fuel pollutants are expected
to increase by at least 3% (CO2Now.org 2011). Although there are projections that by 2030,
emissions in OECD countries will be lessened drastically (ExxonMobil 2010: 34) - these
reductions may not make much difference- on a global scale.

In addition, the world's clarion call is for an uncontaminated source of energy. This explains
why there is a global crave for substitute fuels over conventional fuels like diesel and petrol
(Astbury 2008: 397). Thus, natural gas- in the form of liquefied Natural Gas (LNG hereafter)
is giving the world what it wants. In fact, LNG is non-toxic, clean and green (Kumar et al.
2011: 4265).

Clearly, the call for clean energy, coupled with other factors such as favourable natural gas
prices, growing gas imports, cheaper LNG costs and the crave of gas producers to "cash"
their gas assets- has set the ball rolling for a strong global LNG market. (EIA 2003: v).
Moreover, there are also tons of untapped natural gas resources (Economides and Wood

2009: 1) just waiting to be liquefied. The rest of this paper will be structured follows; section 2 is an overview of the LNG market at the global level, section 3 reviews the LNG market at the regional level, section 4 concludes.

2. AN OVERVIEW OF THE LNG MARKET AT THE GLOBAL LEVEL

There are about 65 liquefaction projects and roughly 181 re-gasification projects that are being built (Kumar et al. 2011: 4098). According to Kumar et al. (2011) IEA projects a quintuple increase in the world's liquefaction facilities by 2030. Besides, Australia and Middle East are set to enjoy huge exports because a large number of liquefaction projects are in those regions (Kumar et al. 2011: 4098).

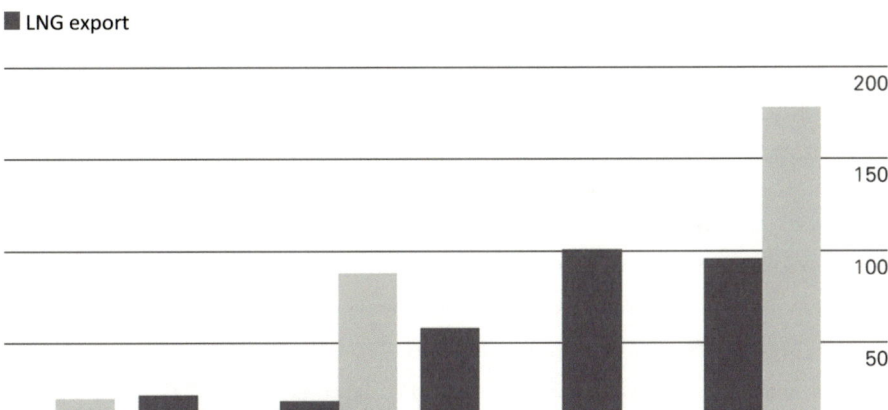

Figure 1: LNG Imports and Exports. (bcm)

Source: (BP Statistical Review of World Energy 2011: 29)

3.   THE LNG MARKET AT THE REGIONAL LEVEL

3.1   THE CURRENT & FUTURE MARKET OF MIDDLE EAST

As shown in Figure 1, the Middle East accounts for the greatest LNG exports on the regional level. Though UAE and Oman have LNG interests this paper will limit itself to Qatar and Iran's status in the Middle East. The reason is that Qatar is a Middle Eastern state that has totally utilized its LNG potential (Economides and Wood 2009: 11). Iran also looks promising; because despite international constraints, it has acquired money to kick start the country's first LNG project (bloomberg.com 2011). It is also worth noting that, UAE exported a total of 23.90 bcm of LNG to Brazil, Kuwait, South Korea, China, Japan and Taiwan in 2010 (BP 2011: 29).

Qatar is the biggest exporter of LNG (Bloomberg.com 2011). In fact, Qatar has production capacity of 77 million tonnes per annum of LNG (Bloomberg.com 2011). Secondly, Qatar is the world's largest LNG exporter- generating over 31mmt per year (Kumar et al. 2011: 4268). Qatar's LNG also has a wide range of importers spanning different continents. Qatar supplied a total of 2.56 bcm of LNG to North America, 0.99 bcm to South and Central America, 30.08 bcm to Europe and Eurasia and 36.2 bcm to Asia Pacific (BP 2011: 29).

Secondly, Iran holds 16 percent of worldwide gas reserves. According to pseez.ir 2010, Iran is a co-sharer of huge gas reserves (South Pars Gas field). Moreover, Iran is also a top gas exporter in the world (bloomberg.com 2011).However; Iran has not been a top exporter of LNG because Iran has technology challenges (iranlng.ir 2011) and has been under international constraints (bloomberg.com 2011). However, Iran stands to be a major exporter of LNG because of its massive gas reserves. According to (iranlng.ir 2011) Iran's South Pars gas reserves – has the capacity to satisfy the world's fuel demand. Surely, China, South Korea and Europe could be a huge market for Iran's LNG export (Omidvar, 2007 cited in Economides and Wood 2009)

3.2   THE CURRENT & FUTURE MARKET OF NORTH AMERICA MARKET

This paper will discuss only the U.S. and Canadian LNG market. The reason is that U.S. was the biggest importer of LNG in the North American market in 2010 (BP 2011: 29). Moreover, consumption of LNG in US is expected to grow (Maxwell and Zhu 2011: 217). Canada is also a huge supplier of natural gas to the U.S. (Economides and Wood 2009: 5). As shown in Figure 1, North America was responsible for importing about 25 bcm of LNG. North America also consumes more LNG than it consumes (Ruester and Neuman 2008: 3160).

According to lngfacts.org 2010, there is a mandate in the U.S. to produce clean energy via LNG. Moreover the U.S. uses natural gas for to power a lot of its activities (Kumar et al. 2011: 4100). Besides, Norway, Egypt, Nigeria, Qatar, and Yemen supplied a total of 12.23 bcm of LNG to the US in 2010 (BP 2011: 29). Trinidad and Tobago also supplies more than 67% of US's LNG imports (Ruester and Neuman 2008: 3163).

LNG also contributes significantly to U.S.'s energy mix especially during strong demands (lngfacts.org 2010). Moreover lngfacts.org 2010, suggests that shale gas- can satisfy U.S.'s energy needs, as well as produce LNG for export. However, this may not be the case because EIA predicts that in the year 2025- US will experience a demand increase of about 76% (Kumar et al. 2011: 4101). Surely, the U.S. will need LNG imports to supplement their energy demands in 2025. In addition to this, global energy needs, favourable natural gas prices and global liquefaction power influences U.S. LNG imports (Kumar et al. 2011: 4101). Thus, U.S. can buy more LNG when there is surplus natural gas on the world market (Kumar et al. 2011: 4101).

Furthermore, Canada is the second biggest producer of natural gas in the Western Hemisphere (Economides and Wood 2009: 5). Moreover, gas corporations in Canada have started looking at constructing LNG receiving terminals (Economides and Wood 2009: 6). These corporations are prepared to either sell LNG to Canada or the US (Economides and Wood 2009: 6). It is worth mentioning that, Canada imported 2 bcm of LNG in 2010; Peru, Norway, Qatar, Trinidad & Tobago were the LNG suppliers (BP 2011: 29).

3.3   THE CURRENT & FUTURE MARKET OF ASIA PACIFIC'S MARKET

According to BP 2011, Asia Pacific- which includes Australia, China, India, Japan, Malaysia, Singapore, South Korea, Taiwan and Thailand- imported about 178 bcm of LNG. This represents approximately 60% of total LNG imports. This paper will limit itself to Japan because Japan is the chief importer of LNG, importing a startling 93.48 bcm (BP 2011: 29). According to BP 2011, Japan was the biggest importer of LNG. According to Figure 1, though Asia Pacific imported 178 bcm (LNG), it also exported about 97 bcm of LNG. Indonesia, Brunei, Indonesia and Malaysia were the LNG exporters in 2010 (BP 2011: 29)

Japan's current huge LNG imports were influenced by the Tohoku earthquake and its related tsunami (eia.gov 2011). In fact, the tsunami and political instability in the Middle East disturbed the energy demand worldwide (BP 2011: 1). In 2010 Japan imported LNG from U.S., Trinidad & Tobago, Belgium, Algeria, Egypt, Equatorial Guinea, Nigeria, Oman United Arab Emirates (UAE), Yemen , Australia, Brunei, Malaysia and Indonesia (BP 2011: 29). Japan will continue to import LNG in the future because it is dependent on gas imports (Economides and Wood 2009: 7). Besides, the strong demand for LNG in Japan has triggered liquefaction investments in Asia (Economides and Wood 2009: 11). Based on this, information – it is safe to predict that Japan would need more LNG to generate electricity especially in the wake of nuclear power disruptions (eia.gov 2011)

Moreover, India and China also have rapid growing economies that will play significant roles- in the LNG market (Economides and Wood 2009: 7). Their annual GDP has been predicted to grow at an average of 6 percent until 2030(ExxonMobil 2010: 8). According to ExxonMobil- China and India's energy needs for industry and electricity are going to grow immensely. It is also worth mentioning that China is now the world's largest energy consumer (BP 2011: 2). China and India currently sources their LNG from Peru, Trinidad & Tobago, Russia, Egypt, Equatorial Guinea, Nigeria, Australia, Indonesia and Malaysia (BP 2011: 29). These countries are set to export more LNG - as China and India grows. According to BP (2011), China and India consumed 24.95 bcm of LNG in 2010.

3.4 THE CURRENT & FUTURE MARKET OF EUROPE AND EURASIA

According to Fig. 1, Europe and Eurasia import about 90 bcm of LNG. This region also exports about 20 bcm of LNG. According to BP (2011) - Belgium, Spain, UK, France, Greece, Italy and Portugal were the LNG importers. However, only Norway and Belgium exported LNG (BP 2011: 29).

In addition to the above, Europe is seriously considering natural gas because of green issues like pollution (Kumar et al. 2011: 4102). There is also wide-spread belief that Europe will become more dependent on natural gas in the future (Kumar et al. 2011: 4102). This future development will be a huge plus for the LNG market in Europe. (Kumar et al. 2011: 4102).

Moreover, Russia is the second biggest natural gas producer (18.5 tcf in 2006) and the largest consumer (Ruester and Neuman 2008: 3160). According to Economides and Wood (2009) Russia controls almost three tenth of the world's gas reserves. However, Russia is not a major player in the LNG market. In fact, According to menafn.com 2011, Russia's Gazprom has two ongoing gas projects – Shtockman and Sakhalin-2., and wants to cash in on the growing LNG market.

3.5 THE CURRENT & FUTURE MARKET OF AFRICA'S MARKET

According to Fig. 1, Africa exported about 58 bcm of LNG but imported none. The exporters were Algeria, Egypt Equatorial Guinea, Libya and Nigeria in 2010 (BP 2011: 29). However, the following countries received imported LNG from Africa in 2010: US, Mexico, Brazil, Chile, Belgium, France, Germany, Italy, Portugal, Spain, Turkey, Kuwait, China, India, Japan, South Korea and Taiwan.

4   CONCLUSION

At the global level, the interest for LNG will not wane but grow stronger- considering the number of LNG projects under construction (Kumar et al. 2011: 4098). Moreover, the Middle East - the leading exporters of LNG are yet to reach their true capacity because of Iran's challenges in monetising its LNG opportunities (bloomberg.com 2011).North America, Europe and Eurasia- will also increase their demand for natural gas- in the light of growing environmental concerns (Kumar et al. 2011: 4102). Russia also has the capacity to huge exporter of LNG, and its pursuing an LNG project now (menafn.com 2011). Although, Asia is currently the top importer of LNG- its demand will grow stronger because China and India are rapid growing countries. (Economides and Wood 2009: 7). As a matter of fact, even the price of LNG will be influenced largely by China and India's exports (Lin et al. 2010 : 4390) Thus, the future market of LNG will be determined by Asia and Middle East.

REFERENCES

Astbury, G.R (2008) 'A review of the properties and hazards of some alternative fuels'. *Journal of Process Safety and Environment Protection* 86 (2008), 397-414

Bloomberg (2011) Iran LNG Says It Will Overcome Sanctions to Start Exporting Fuel In 2012 (online) available from < http://www.bloomberg.com/news/2011-04-13/iran-lng-says-it-will-overcome-sanctions-to-start-exporting-fuel-in-2012.html>

British Petroleum PLC (2011) BP Statistical Review of World Energy June 2011. London: BP Statistical Review Printing Office

Center for Liquefied Natural Gas (2010) LNG Future (online) available from <http://www.lngfacts.org/LNG-Future/default.asp>

CO2Now.org (2011) What the Earth Needs to Watch (online) available from <http://co2now.org/>

Economides, M. J. And Wood, A. W. (2009) 'The State of Natural Gas'. *Journal of Natural Gas Science and Engineering* 1 (2009) 1-13

EIA (2011) Japanese Power Companies using more LNG to Generate Electricity (online) available from < http://www.eia.gov/todayinenergy/detail.cfm?id=2430>

Energy Information Administration (2003) *The Global Liquefied Natural Gas Market: Status and Outlook.* Washington DC: EIA Printing Office

ExxonMobil (2011) *2010 The Outlook for Energy: A View to 2030*. Texas: ExxonMobil Printing Office

Iran LNG (2011) Iran and LNG market (online) available from<http://www.iranlng.ir/article-en-20.html >

Kumar, S., Kwon H., Choi K.,Lim W., Cho J.H., Tak K., and Moon I. (2011) 'LNG: An eco-friendly cryogenic fuel for sustainable development'. *Journal of Applied Energy* 88 (2011), 4264-4273

Kumar, S., Kwon H., Choi K., Cho J. H., Lim W. and Moon I. (2011) 'Current Status and Future Projections of LNG Demand and Supplies: A Global Prospective'. *Journal of Energy Policy* 39 (2011), 4097-4104

Lin, W., Zhang, N., and Gu A. (2010) 'LNG (Liquefied Natural Gas): A Necessary Part in China's Future Energy Infrastructure'. *Journal of Energy* 35 (2010), 4383-4391

Maxwell, D. and Zhu, Z. (2011) 'Natural Gas Prices, LNG transport costs, and the Dynamics of LNG Import'. *Journal of Energy Economics* 33(2011) , 217-226

Menafn.com (2011) Russia- Gazprom to Jump on Booming LNG Market by 2030 (online) available from < http://www.menafn.com/qn_news_story_s.asp?StoryId=1093456855>